VEHICLES
ON THE JOB

FOUR-TRACK TRACTORS
ON THE FARM

by Samantha Bell

Cherry Lake Press

Published in the United States of America
by Cherry Lake Publishing
Ann Arbor, Michigan
www.cherrylakepublishing.com

Reading Adviser: Beth Walker Gambro, MS, Ed., Reading Consultant, Yorkville, IL
Content Adviser: Louis Teel, Professor of Heavy Equipment Operating, Central Arizona College

Photo Credits: © dvande/Shutterstock, © Jovanovic Dejan/Shutterstock, cover, title page; © dvande/Shutterstock, 4; © clynt Garnham Agriculture/Alamy Stock Photo, 6; © Zigmunds Dizgalvis/Shutterstock, 8; © digitalreflections/Shutterstock, 10; © marchello74/Shutterstock, 12; © David Wright/Flickr, 14; © BG Media/Shutterstock, 16; © Zoran Zeremski/Shutterstock, 18; © Rick Dalton - Ag/Alamy Stock Photo, 20

Cherry Lake Press is an imprint of Cherry Lake Publishing Group.

Library of Congress Cataloging-in-Publication Data has been filed and is available at catalog.loc.gov

Cherry Lake Press would like to acknowledge the work of the Partnership for 21st Century Learning, a Network of Battelle for Kids. Please visit Battelle for Kids online for more information.

Printed in the United States of America

Note from Publisher: Websites change regularly, and their future contents are outside of our control. Supervise children when conducting any recommended online searches for extended learning opportunities.

TABLE OF CONTENTS

JOHN
9RX
590

TRACK POWER

Some tractors do not have rubber tires. They have 4 rubber **tracks**.

CASE

Tracks give the tractor better **traction**. They help the tractor move over soft ground.

THINK

What kinds of soft ground could a four-track tractor move over?

8

GET A GRIP!

The tractor goes up hills.
The tracks keep it balanced.
They can grip the ground.

CASE
QUADTRAC
500

The tracks move on soft soil.
They go over rough ground.

LOOK

See how the tracks wrap around wheels? How do you think the wheels help move the track?

FIELD WORK

A tractor works in a field.
It pulls a tiller. It **tills**
the ground in straight rows.

450

The tracks keep the ground even.
The ground will not get **ruts**.

CASE IH
600

The soil is now ready for planting.

CONNECT

What are other things plants need to grow?

HARVEST HELP

The farmer plants the seeds. Soon they will grow into **crops**.

1050
CASE IH
450

The tractor can pull a grain cart to take the crops to a truck. The grain can be made into bread and other foods. Yum!

FIND OUT MORE

Go online with an adult to find videos of four-track tractors. Write down some things you **see** and **hear**.

Then, write what you **think** about those things. **Make a connection** to something else you know.

Next, write what you **wonder** about four-track tractors. from the videos. What would you like to learn more about?

GLOSSARY

crops (KRAHPS) plants that are grown as food

ruts (RUHTS) places in the ground worn down by movement over time

tills (TILLZ) breaks up soil using a piece of equipment pulled by a tractor

tracks (TRAKS) the metal or rubber belts a vehicle uses to move

traction (TRAK-shuhn) the force that causes a moving thing to grip a surface

INDEX

ABOUT THE AUTHOR

Samantha Bell has written over 100 books for children. She lives in South Carolina with her family and pets.